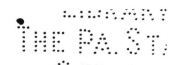

MACHINERY'S REFERENCE SERIES

EACH NUMBER IS ONE UNIT IN A COMPLETE
LIBRARY OF MACHINE DESIGN AND SHOP
PRACTICE REVISED AND REPUB-
LISHED FROM MACHINERY

NUMBER 38

GRINDING AND LAPPING

CONTENTS

Influence of Vibrations on Action of Grinding Machines.

The grinding wheel rotating at a high-speed tends to jar its bearings and supports. Vibrations of this kind would result in an oscillating motion of the grinding wheel perpendicular to its own axis of rotation and along the line connecting the center of work with the center of the wheel. The frequency of these vibrations depends entirely upon the weight of the oscillating parts. The cause of the vibrations is that the center of gravity of the rotating parts, grinding wheel, shaft, pulley, etc., is not entirely the same as the center line of rotation. This is partly due to the uneven structure of the material. It is very plain to everybody, that the oscillating grinding wheel cannot cut to its full capacity. The length of the oscillations

TABLE I. SPEED OF EMERY WHEELS.

Diameter of Wheel, inches.	R. P. M. for Surface Speed of 5,000 feet	R. P. M. for Surface Speed of 6,000 feet
1	19,099	22,918
2	9,549	11,459
4	4,775	5,730
6	3,183	3,820
8	2,387	2,865
10	1,910	2,292
12	1,592	1,910
14	1,364	1,637
16	1,194	1,432
18	1,061	1,273
20	955	1,146
24	796	955
30	637	764

might not be large, perhaps only one-thousandth of an inch or a fraction thereof, but the cut will be just so much deeper one moment than the next following. Only at one moment, when the wheel is furthest in, will it cut to its full capacity.

It is very important, in order to secure nice running of the wheel, to have the belts in good order, and to have the boxes closely adjusted, even though they run a trifle warm. Because of the high speed of the shaft, the boxes ought to be made with ring-oiling devices. This would allow a closer adjustment, and secure a better running of the shaft. However, as far as the writer knows, there are no grinding machines on the market equipped with ring-oiling boxes. The slides should, for the same reason as the boxes, be adjusted closely, even though they slide hard.

Speed of Grinding Wheels.

The peripheral speed of the grinding wheel should be approximately from 5,000 to 5,500 feet per minute. There are occasionally cases when higher speed is desirable, but with higher speed there is danger of the wheel breaking. The wheel should, however, never be run slower than 5,000 feet per minute, because it becomes less efficient at slower speeds.

Above will be found a table which gives the number of revolutions per minute for specified diameters of wheels to cause them to run at the respective periphery rates of 5,000 and 6,000 feet per minute.

Experience has shown that for grinding work with fairly large diameter, better results are obtained by using a comparatively small wheel than by using one with too large a diameter. The explanation of this fact is that the wheel of smaller diameter clears itself faster from the work, while the larger one has a larger contact surface, and, therefore, the specific pressure between wheel and work becomes reduced, and the metal removed by the wheel stays too long a time between the wheel and work; and prevents the particles of the wheel from cutting properly into the work. The peripheral speed must, however, be the same for the smaller wheel as for the larger one.

Surface Speed of Work.

The proper surface speed of the work varies somewhat with the material and kind of work to be done. The grinding machine builders recommend 15 to 30 feet as a good average speed range for ordinary kind of work. For cast iron this can be slightly increased. The writer has had experience in grinding a very tough and hard steel (manganese steel), and has found the right surface speed in this special case to be as low as 6 to 8 feet a minute for rough grinding. For finishing grinding, the speed should be somewhat higher than for rough grinding. For delicate work the speed should be slow, because the work could easily be damaged by forced grinding.

As a general rule, for determining the surface speed for a certain kind of material, one can say that a brittle material, as cast iron, takes a high speed, while a tough and hard material, as the best tool steel, takes a slow speed. For grinding close to size and for high finish, the depth of the cut must be small, and higher surface speed can consequently be used.

Many of the grinding machines on the market are built so as to have the work revolving on two dead centers. This is done more for the sake of being able to obtain accuracy than for the sake of increasing the cutting efficiency of the machine.

Traverse Speed of Grinding Wheel.

The traverse speed of the grinding wheel should for ordinary grinding be three-fourths of the width of the wheel, that is, for one revolution of the work the wheel should travel three-fourths of the width of the face. If the wheel be traversed slower, the new cut is overlapping the old one more than necessary, and too large a part of the wheel is idle. It is, however, necessary that the new cut overlaps the old one with about one-fourth of the width of the face, because the edges of the face easily become rounded off, and, if the travel be too rapid, the result is an uneven surface.

The capacity of the wheel, within certain limits, of course, is proportional to the width of the face. A certain specific pressure between wheel and work is required for the highest cutting capacity. A wider wheel requires consequently a larger total pressure. But many of the machines now on the market are not rigid nor heavy enough to stand the pressure needed for a fairly wide wheel, cutting at full load, without vibration and chatter. The grinding machines on the market have not, in the writer's opinion, yet reached their full capacity. Wider wheels should be used, and the machines be designed and built heavier in order to take the load of the cutting wheel, without perceptible vibration of the machine.

For the final smooth finish, a slower traverse speed should be used, especially if the face of the wheel is not kept a perfectly straight line. A smoother surface is obtained by using a slower traverse speed. The part of the wheel which is overlapping, while theoretically it does not cut, still wears away the unevenness left from the first cut, and, by this, to some extent polishes the surfaces.

While grinding a plain cylindrical piece of work, the grinding wheel should not be allowed to travel too far past the ends of the piece before reversing; it is only waste of time. The wheel should be reversed when three-fourths of its width is past the end of the work.

Depth of Cut.

The depth of the cut to be taken depends upon the material, kind of wheel, and the work done. It should be deep enough to permit the wheel to do its utmost. This is, of course, true only about pieces that are rigid enough to stand a heavy cut. The grinding operator himself will have to determine the depth of cut for each individual case, judging it by the prevailing conditions of work, machine, and wheel.

When the piece to be ground, owing to the hardness of the material, cannot be roughly finished by a cutting tool before being placed in the grinding machine for the final finish, there is often up to 3/16 inch on the diameter to be removed by grinding. Employing the same principle as when the piece is previously to the grinding operation roughly turned in a lathe, the work should first be put up in a machine equipped with a coarse and wide grinding wheel. A wheel of this kind is capable of removing stock rapidly. The piece should be finished to within 0.005 inch of the finished diameter in this machine, and then moved to a machine equipped with a finer grain wheel, and the final finish given to it.

The Grinding Wheel.

For heavy grinding, the alundum wheel is the best for removing stock rapidly. The carborundum wheel will give a smoother finish, and is to be recommended for the large majority of other classes of grinding. Emery is less abrasive, but gives a higher polish. Most grinding wheel manufacturers recommend their medium grade, M.

The question as to what is the very best wheel for finishing any particular piece cannot be definitely answered. On the next page is given a table of wheels which can with advantage be used in the cases mentioned. This table is recommended by one of the largest grinding machine manufacturers.

Grit No. 24 may be too coarse for any but rough classes of work, but if mixed with No. 36 it gives a fair result. No. 30 used separately is capable of a very fair commercial finish, but if mixed with No. 46 will give as fine a finish as is desired by the majority of the grinding machine users, and at the same time it retains the rapid cutting capacity. Nos. 46 and 60 are as fine as is necessary for almost any manufacture, although finer than these are used by some concerns who require a very high gloss finish.

A satisfactory grinding wheel is an important factor in the production of good work. In machine grinding, it is desirable, in order that the cut may be constant, and give the least possible pressure and heat, to break away the particles of the wheel after they have become dulled by the act of grinding. It is the faculty of yielding to or resisting the breaking out of the particles which is called grade. The wheel from which the particles can be easily broken out is called soft, and the one that retains its particles longer is called hard. It is evident that the longer the particles are retained the duller they will become, and the more pressure will be required to make the wheel cut. Retaining the particles too long causes what is familiarly known as glazing. A wheel should cut with the least possible pressure, to effect which it must always be sharp. This is maintained by the breaking out of

TABLE II. GRADE OF WHEEL TO USE FOR DIFFERENT MATERIALS.

Material	● Grit No.	Grade
Soft Steel { Ordinary shafts....	24 to 60	Medium.
Soft Steel { Steel tubing or very light shafts......	24 to 60	Two or three grades softer than medium.
Tool steel or cast iron.....	24 to 60	Medium or one grade softer.
Internal grinding	30 to 36	Medium or several grades softer.

particles. Therefore, a wheel of proper grade cutting at a given speed of the work possesses "sizing power," or ability to reduce its size uniformly without breaking away its own particles too rapidly; obviously if the work is revolved at a higher speed, the particles will be torn away too fast, and the wheel will lose its sizing power.

The properties of toughness and hardness of the material to be ground has a retarding influence on the grinding because it makes the material stick to or clog the wheel. The ground-off material, instead of being thrown away from the wheel by the centrifugal force, gets in between the particles of the grinding wheel. It is self-evident that this has a greatly retarding effect on the cutting quality of the wheel. A brittle material, on the contrary, does not have the tendency of clogging the wheel, but the stock ground off is immediately thrown away from the wheel, leaving the particles free to cut without the retarding action of undue friction, and generation of more than the due amount of heat. If we take into consideration only these properties of the material to be ground, the tough or leady material requires a soft wheel, because the particles must break away ·fast enough to prevent the wheel from being clogged. In this case, the particles do not wear enough to become dull, but must break away before this. When grinding a brittle or hard material, on the contrary, the wheel is less liable to be clogged, the particles do not need to break away so soon, and, therefore, a harder wheel should be used. However, the wheel must not be so hard that the particles get too dull and become inefficient as cutting agents before they break away.

Importance of Wheel Running True.

In order to obtain the full efficiency of the grinding wheel, it must be run perfectly true; that is, cut evenly all the way around. The grinding wheel detects its own errors. A slight difference in the

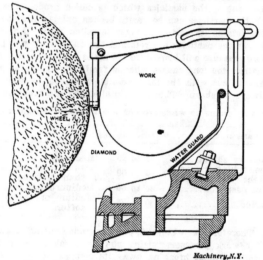

Machinery,N.Y.

Fig. 1.   Fixture for Truing Emery Wheel with Diamond.

sparks indicates that the wheel is out of true. The eccentric wheel has about the same kind of action as the one which is vibrating because of too weak supports. Furthermore, the edge of the grinding wheel should be kept perfectly straight. If the edge be curved, however slightly, a curved cut will be the consequence. Many grinding machines give inefficient results because the edge of the wheel is not kept in a true straight line. The operator seldom appreciates the great importance of this, and, therefore, the foreman should watch the men closely in regard to this point.

The best tool for truing the wheel is the diamond, but, this being rather expensive for shops where not very much grinding is done, the usual emery wheel dresser can be used to good advantage. In truing the wheel, the dressing tool should be kept stationary and rigidly supported, and the wheel should be traversed back and forth, until a true edge is obtained. Fig. 1 shows a fixture and arrangement for wheel truing with a diamond.

Wet and Dry External Grinding.

Nearly all plain cylindrical grinding is now done wet. There are many reasons why the wet method is to be preferred to the dry. Because of the friction between the grinding wheel particles and the work, as well as between the cut-off material before it leaves the wheel and the work, more or less heat is generated. If this heat is not carried away, the work will be burned. Besides, the edge of the grinding wheel would be highly heated, but the center would still remain comparatively cool, and the outside would expand and there would be danger of the wheel breaking. It is found that the water has a softening effect upon the wheel, therefore a harder wheel is required for wet grinding than for dry.

Machines with Two Grinding Wheels.

The grinding machines on the market are equipped with only one grinding wheel, but there is no reason why two grinding wheels cannot be employed to advantage. In this case one wheel is to operate on each side of the work. As both of the wheels are to throw the sparks and the water down, one of the wheels has to cut with the revolving of the work, that is, the peripheries of the wheel and the work are going downwards. This is, of course, not the ideal condition, but, when the work is revolving at a slow peripheral speed, there is not much difference in the cutting capacity of the two wheels.

It is self-evident that, when employing two wheels, one at each side of the work and just opposite each other, the traverse speed of the wheels must be twice as fast as in the case of only one wheel, or three-fourths of the width of the wheel for one-half revolution of the work. Otherwise one wheel will overlap the cut of the other.

The two machines for external grinding which the writer designed, have two wheels working according to the principle previously described. Fig. 2 gives an idea of the arrangement used on one of these machines. The principal features of the design can be studied direct from the illustration without any further comments.

One new feature of these machines is that each grinding wheel is driven independently by a motor. This motor is mounted above the wheel spindle, and is belted directly to same. Special attention has been paid to designing the support of the motor in order to prevent the vibrations of the motor from being transferred to the grinding wheel.

Internal Grinding.

The development of internal grinding is not, by far, so advanced as that of external grinding. To be sure, there are a few machines and fixtures on the market that are designed and built for the internal grinding of holes of various kinds, but the machines suffer from lack of rigidity, and some of the most conscientious grinding machine builders do not recommend them very highly, but admit their inefficiency for removing any comparatively large amount of stock. It has even gone so far that one man holding a prominent position with one of the largest grinding machine manufacturing concerns in the country has said, that in his opinion, the internal grinding machine is a mistake from start to finish, and that it will never be made a success.

This state of affairs has not come about without good reason. As we already have seen, the rigidity of the arrangement for supporting the grinding wheel is a very important factor for all efficient grinding.

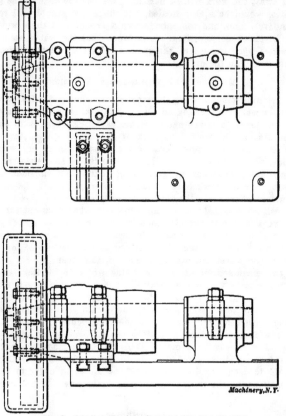

Fig. 2. Grinding Head for External Grinding Machine.

But the internal grinding machine does not very well lend itself to the employment of any rigid and heavy fixtures, and the grinding wheel must necessarily be small, and therefore lacks the strength to stand a heavy cut. The designer, when designing the fixtures for internal grinding, has an entirely different problem to solve than when designing those for external grinding, where it is comparatively easy to obtain ample rigidity. The internal grinding wheel must be mounted at the end of a small spindle which projects past the bearing far enough to enable the wheel to reach past the end of the hole to be ground. Such a spindle rotating at a high speed is liable to vibrate, especially if pressure be applied at the end of it, as is here the case.

Sometimes, however, it becomes absolutely necessary to grind, internally, even a comparatively large amount of stock. This is the case, when finishing manganese steel, this material being so hard that it cannot be cut by any kind of tool steel. Take the case of bores of manganese steel car wheels. As the grinding of the bores must be done without any stock having previously been removed from the rough casting, on the average about one-eighth inch of metal must be ground off from the hole. All the errors in the cored hole, as eccentricity in reference to the circumference of the wheel, etc., must be corrected by grinding. A hole cored in a manganese steel casting is always comparatively much rougher than a hole cored in cast iron, and all this must be taken into consideration, when determining the amount of stock to leave for the grinding process.

Design of Heads for Internal Grinding.

The fixture used in the internal grinding machine designed for grinding these wheels is shown in Fig. 3. Internal grinding fixtures gen-

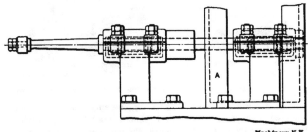

Machinery,N.Y.

Fig. 3. Grinding Head for Internal Grinding.

erally have a long extension bearing, as shown in Fig. 4. This serves to support the spindle as near to the grinding wheel as possible; but the diameter at the root of this extension, that is, nearest to the box, cannot exceed the diameter of the grinding wheel.

The spindle, shown in Fig. 3, is made solid, and has the largest diameter possible for the size of the grinding wheel. An increased amount of rigidity and a greatly increased simplicity is gained by this design.

When working, the grinding wheel produces, especially in dry grinding, very much dust. When inside a hole the dust cannot very easily get away, but whirls about in the hole. If the spindle has a bearing near to the grinding wheel, the dust will find its way into the journal. This drawback is entirely eliminated by having a large solid spindle without a bearing near to the grinding wheel.

As to the relation between the overhanging part of the spindle and the distance between centers of the boxes, there are many factors that come into consideration in regard to this relation, such as the design of the boxes, the diameter of the spindle, how close the spindle can be allowed to run in the boxes, etc. However, the distance between the centers of the boxes should be made as large as the general design conveniently permits.

Fig. 3 shows at *A* the support for the motor. This support is placed on the top of the top rest. The driving pulley is placed between the bearings, so that the support could be made as rigid as possible.

It was found by actual experience with these fixtures, that when the grinding wheel was taking a fairly heavy cut the spindle did not vibrate nearly so much as when the wheel was running idle. The springing quality of the spindle, and the pressure between work and wheel, made the wheel cut without any chattering worth mentioning.

Regarding the peripheral speed of the grinding wheel, what has already been said with reference to external grinding is equally applicable to internal grinding.

Because of the lighter fixtures, the speed of the work should be slower than for external grinding. The writer has found the right cutting speed for hard and tough steel to be, for heavy grinding, about

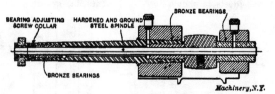

Fig. 4. Common Construction of Grinding Heads for Internal Grinding.

seven feet a minute. For the finishing, the speed can, with advantage, be somewhat higher. The wheel should travel three-fourths of its width for one revolution of the work, the same as for external grinding.

Wet and Dry Internal Grinding.

One point that has been much discussed in regard to internal grinding, is whether it shall be conducted wet or dry. Some grinding machine designers have advanced the opinion that it, by all means, must be done dry, but others claim the wet method to be superior. For light finishing grinding one method might be considered as good as the other, because so small an amount of heat is generated that there is no danger of burning the material or breaking the wheel. But, for heavier grinding, a considerable amount of heat is generated, and it becomes necessary to carry it off by water. At least, such is the writer's own experience on this subject. At a test recently conducted to find out the actual difference between dry and wet internal grinding, it was found that the cutting quality of the grinding wheel was about the same in both cases, but, with a heavy feed and dry grinding, the work was highly heated, and the wheel broke after about half an hour's run, while, with wet grinding, the wheel stood the heavy cut continuously without breaking.

The water can be injected into the hole in a stream about 1/16 inch in diameter. In addition to carrying away the heat, the water serves to wash away the removed stock from the hole.

Tests have been undertaken on the above mentioned internal grinding machines, in order to find out the time required to grind the bores of a certain kind of manganese steel car wheels. Two different kinds of wheels were tested. The first one, a 20-inch diameter wheel, had a bore 2⅞ inches in diameter and 5⅞ inches long, and it was to be ground for a press fit. The second one, an 18-inch diameter wheel, had a bore 3¼ inches in diameter and 4½ inches long, and was also to be ground for a press fit. Four wheels of each kind were ground during the course of the test, and it was found that the actual time for the grinding operation, not including the time required for putting up the work in the machine, was, for the first kind of wheels 1 hour and 23 minutes for all four, and for the second kind, 1 hour and 9 minutes for four wheels. Considering that the bores of the wheels were not previously turned, but entirely rough, as the wheels were taken directly from the foundry, and considering the hardness and toughness of the steel, the results obtained were considered good. The time of putting up the work in the machine was about 6 to 8 minutes for each wheel. As the machines work automatically, one man is able to run three machines. Counting 8 minutes for the putting up of each wheel, the man is able to grind one wheel of the first kind in 30 minutes, and one wheel of the second kind in 26 minutes.

The work was revolved at a speed of 7.7 revolutions per minute. This makes a peripheral speed, for the first case, of 5.8 feet per minute, and for the second case, of 6.6 feet per minute. The grinding wheel used was a 2-inch diameter, 1-inch face, No. 46 grit, O grade alundum wheel. It was run at a speed of 4,750 feet per minute.

The traverse speed of the work was as high as 0.84 inch per revolution of work. This allowed the wheel to overlap the old cut by only 0.16 inch, but, as the grinding wheel was trued very carefully, this was found to be all that was required for obtaining a nice smooth surface. The traverse feed was not slowed down, but remained the same while doing the final finishing, and a very satisfactory finished hole was obtained. The test was made throughout with wet grinding.

For heavy cylindrical grinding, which has especially been referred to, the width of the wheel used varies between 1½ and 2½ inches, regardless of the diameter. In some special cases narrower wheels than 1½ inch are used, but these special cases are exceptions to the general practice, and must be recognized as such by the machine builders and users. Although larger wheels are used, there is no doubt that the best range of diameters of wheels is between 12 and 18 inches. For how wide a wheel the grinding machine in the future can be designed, has yet to be decided; but, wider wheels and heavier machines point the direction of the road which the designer and machine builder should follow for the development of the grinding machine.

CHAPTER II.

THE DISK GRINDER.

In any machine shop or department of a manufacturing plant where tools for manufacturing operations are made, a properly designed and equipped disk grinder should be considered almost indispensable. For a large portion of the operations most commonly done with a file, and many that are considered surface grinder, milling machine or shaper jobs, can be done better and quicker, and at less cost for files, cutters, etc., with a disk grinder.

As a simple example we will take the case of a piece of tool steel needed, say for a box tool, a back rest, a cutter, or a forming tool, to be say, ¼ inch thick, 1 inch wide, 2 inches long, ends and sides straight and square all around. Probably the bar steel ¼ inch × 1 inch will be enough oversize to grind on a disk to exact size, but not enough oversize to work with a milling machine, shaper or surface grinder. Even if larger stock, say 5-16 inch × 1⅛ inch, or a forging, is used,

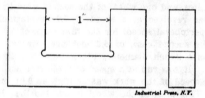

Fig. 5. Snap Gage Finished on Disk Grinder.

it is only necessary to rough one flat side and one edge down fairly close to size and finish all over on a disk grinder. For squaring the ends of one piece like this, and bringing it to exact length, the saving in time over the common way is considerable. Suppose this piece has to be hardened, and after hardening must fit a certain space. It will need truing up after hardening, and here again the disk grinder proves its adaptability.

Regarding the degree of accuracy obtainable with a disk grinder, an example may be of interest. An experienced tool maker was with an exhibit of disk grinders at a fair. Having plenty of time on his hands, he employed a part of it in grinding up six steel pieces, each a one-inch cube. He got the pieces planed roughly in the bar, a little oversize, and sawed off a little long. On his spare time he ground them to one-inch cubes, measuring them with a 1-inch micrometer caliper. When he had finished with them there was no point on any of the cubes that varied more than 0.00025 inch from 1 inch. Packing them together with any combination of sides, the greatest variation from 6 inches as measured with a 6-inch micrometer was 0.0005 inch. All the sides of all the cubes were so nearly square with each other that no error could be detected with a hardened steel square. In grinding these cubes no fixture or clamp of any kind was used. They were laid on the swinging table, and against the rib and pressed against the wheel with the fingers.

A few examples of the application of the disk grinder to tool-room work will give a general idea of its application. Suppose a snap gage such as shown in Fig. 5 is to be made. With the disk grinder the

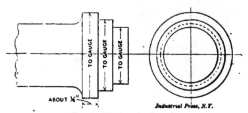

Fig. 6. Work to be Gaged.

gage can be finished all over, sides, edges and ends, and corners beveled or rounded. In hardening the gage springs somewhat, but can easily be squared again on the disk grinder. We are now ready to grind the notch to size. Lay the piece on the swinging table, with the back edge against the rib, the wheel being in the notch. The piece is now ground on both sides without turning it over. This will make the faces of the notch parallel with each other, which they might not be if the piece were turned over. By the use of an end measure gage the snap gage in Fig. 5 is now easily completed.

In a certain shop a job came up to be done in the turret machine.

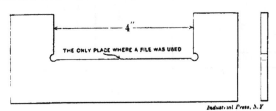

Fig. 7. Snap Gage Ground to Size on Disk Grinder.

A number of cast iron pieces, of the shape shown in Fig. 6, were to be machined. There were eight different sizes of pieces and three dimensions made to gage on each piece, making 24 dimensions in all. The largest dimension on the largest piece was about four inches. The smallest dimension on the smallest piece was about ¾ inch. A few thousand pieces of each size were to be made. Extreme accuracy was not required; a variation of 0.001 inch was allowable. A tool-maker was given the job of making a set of snap gages.

Taking the figures, he made 24 end measure pieces from 5-16-inch round drill rod, hardened them, and marked the size. He then cut the gages from ⅛-inch thick sheet steel, as shown in Fig. 7. The working faces were hardened and ground to the end measure pieces on the disk grinder, and the edges squared and the corners rounded in the same machine. The gages were not touched with a file except to smooth off the edge in the bottom of the notch.

The examples given indicate the use of the disk grinder as a tool-

Fig. 8. Form of Hollow Cast Iron Block used for Test.

room machine. This machine, however, is also efficient for removing large amounts of metal in short time. The efficiency of the machine for this purpose depends largely upon the kinds of disks used. Tests were made at the shops of the Gardner Machine Co., Beloit, Wis., to determine the comparative efficiency for grinding cast iron, of differ-

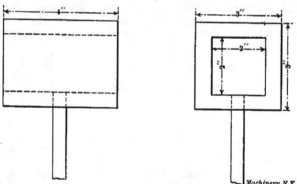

Fig. 9. Hollow Cast Iron Block used for Test.

ent kinds and makes of disks, such as are commonly used in connection with disk grinders. In the following table the different kinds of disks are indicated by figures:

No. 1 indicates the Gardner improved abrasive disk No. 126. No. 5 is the regular No. 24 commercial emery cloth. No. 6 is the same in emery paper. Nos. 2, 3, and 4 are disks of excellent quality as compared to commercial emery cloth.

The disks tested were all 20 inches in diameter and all excepting Nos. 5 and 6 were No. 16 grain. The grinding was done on the ends of hollow blocks of cast iron, as shown in Figs. 8 and 9. The area ground at the end of blocks was 5 square inches. Reducing the blocks one inch in length indicated the removal of 5 cubic inches of metal. The grinding was all done on the same machine by the same operator.

The micrometer stop at the back of the table was set to grind off a fixed amount, usually 0.050 inch, and the twelve blocks ground to the stop. The stop was then moved back 0.050 inch and the operation repeated until the blocks became too warm for efficient grinding, when

TABLE III. RESULTS OF TEST OF EFFICIENCY OF ABRASIVE DISKS.

Disk Number.	Time Used in Minutes.	Stock Removed in cubic inches.	Number of Times Dressed.	Average Cutting Rate, cubic inches per min.	Cutting Rate, First Half of Time Used.	Cutting Rate, Second Half of Time Used.	Life of Disk, Based on Disk No. 1.	Stock Removed, Based on Disk No. 1.
1	754	349.85	0	0.464	0.442	0.486	100. %	100.0%
2	187	42.13	6	0.307	0.344	0.270	18.1%	12.4%
3	540	113.95	0	0.211	0.238	0.184	71.6%	32.3%
4	68	27.97	2	0.411	0.546	0.276	9.0%	8.0%
5	71	2.41	4	0.034	0.062	0.006	9.4%	0.7%
6	73	12.48	2	0.171	0.273	0 069	9.7%	3.5%

they were cooled, and the time of grinding and the amount of metal removed, noted. This was repeated until the disk was worn out or the blocks all ground up. In the latter case, new blocks were substituted and the operation continued until the disk was worn out. By reversing the blocks they were ground down until the wheel touched the handles on both sides. During this test several hundred pounds of these blocks were converted into cast iron chips.

It will be noted in Table III, that it was necessary to use a Huntington emery wheel dresser on all disks tested except Nos. 1 and 3. The dresser was used whenever the surface of the disk became dull and glazed so that it would not cut cast iron readily. The use of a dresser shortens the life of the disk, but it is absolutely necessary.

CHAPTER III.

COST OF GRINDING.

To figure, with any degree of accuracy, the cost of commercial wet grinding, requires considerable experience in the use and management of the machine, in order to be as closely approximated as lathe work. There also seems to be a greater difference in operators, due partly no doubt to the fact that the general use of the grinder has not yet become as common. A great many operators seem to be afraid to push their machines, and spend a good deal of time in useless calipering. They seem to forget that if they have several thousandths to take off a piece and are feeding in one or two thousandths at each reversal of the machine, they need not caliper until within one or two thousandths of size, if they will keep in mind the number of reversals the machine has made. And another class seem to think that because grinding is a finishing job, it must be nursed.

As a matter of fact there is no machine which so readily and accurately responds to the touches of an operator as the wet grinding machine. Of course there are delicate pieces and certain shapes which

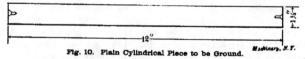

Fig. 10. Plain Cylindrical Piece to be Ground.

Machinery, N. Y.

have to be carefully handled, but the usual run of work is so simple that any good apprentice can be put on it and taught in a short time.

As the work usually comes from the lathe, with approximately 1/64 to 1/32 inch stock to be removed, a few reversals of the machine with the work taking nearly the full width of the wheel each revolution and a cut of two to four thousandths until nearly up to size and then a much slower traverse per revolution for finishing, according to the kind of finish desired, and the work is done. To obtain the best speed, the limits required on the lathe must not be made too narrow, from 1/64 to 1/32 inch being admissible for ordinary work, and more on large work; for the facility of the grinder in finishing work is far in excess of the lathe, and the latter must be relieved of all the finishing possible.

To figure the actual time for removing stock on the grinder we must take into account the longitudinal traverse of the wheel for each revolution of the work, the surface speed of the work and the depth of the cut. The latter must be varied according to the nature of the material, greater or less according to whether it is hard or soft; and the traverse per revolution of work is lessened if a fine finish is desired. The shape of the piece also somewhat affects both of these points, as long, thin pieces require a slower traverse and lighter cuts.

Take, for instance, the plain piece, Fig. 10; material, hardened steel. For this a work surface speed of 15 feet, or about 37 revolutions per minute would be suitable. Assuming we have a wheel 18 inches in diameter, and 1½ inch face, a traverse of two-thirds the face of the wheel or one inch per revolution of work is usual. This would require 12 revolutions to pass the length of the piece, plus 1 revolution for clearance, or for dwell if there happens to be a shoulder. This would make, roughly, three reversals a minute.

On a medium-sized machine an automatic feed equivalent to a work reduction of about 0.002 inch would be suitable, or a reduction of about 0.006 inch per minute. If the work came with an average allowance of 0.030 inch for grinding, it would require theoretically 5 minutes actual grinding time to rough this piece down. To this must be added the time for handling the work, adjusting the machine and back rests (in this case only one rest would be used), calipering the work and finishing. This time will amount to as much as the grinding time with most operators (most of it being taken up in finishing), which would make the actual time about ten minutes apiece. As a matter of fact, work of this size is actually being ground at the rate of seven or eight pieces per hour.

If a fine finish is desired a higher work speed and slower traverse would be required. For a very fine finish a work speed of 45 feet surface speed and traverse of 1/6 inch per revolution would be suitable for finishing, with, of course, a very much smaller feed. This change in the work and traverse speed could be made when the work is nearly up to size, and would probably require about three minutes. If the piece were of soft steel a deeper cut can be taken and a wider traverse, a cut of 0.003 inch and a traverse nearly up to the width of the wheel being admissible. In grinding long shafts it is necessary to allow proportionately more time for adjusting back rests and for calipering, to insure that the piece be straight. This often takes twice the actual grinding time.

Now let us look at the more complicated piece, Fig. 11. This will have to be done on a larger machine, and the larger machines are slower to handle. This piece is a piston rod of 40 carbon

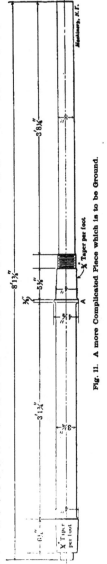

Fig. 11. A more Complicated Piece which is to be Ground.

steel. We will use for this a 20-inch wheel of 2½ inches face. A suitable traverse for this would be 2 inches per revolution and a surface speed of 15 feet would make about 19 revolutions for the part 3 inches in diameter, and about 15 for the part 3¾ inches. The figures would be about as follows:

Total amount to be removed, 0.060 inch; amount per reversal, 0.004 inch; number of reversals required, 15.

3 inches diameter, to cross once, 1 1/5 minute; total for 15 reversals	18 minutes
3¾ inches diameter, to cross once, 1½ minute; total for 15 reversals	22 minutes
Tapers, both, to cross once, 2/5 minute; total for 15 reversals	6 minutes
Setting up and adjusting	10 minutes
Total	56 minutes

If it be desired to put a radius on the wheel and grind the fillets at shoulder *A*, about 10 minutes more should be allowed; and if there

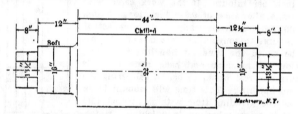

Fig. 12. Chilled Cast Iron Roll Ground from the Rough.

were more than one piece to be done considerable time could be saved in setting for the tapers.

The piece, Fig. 12, is a chilled cast-iron roll to be ground from the rough. This will take the largest machine built, and here the time taken is almost all grinding time. The average reduction for the chilled part is ¼ inch, and a feed of 0.002 inch is about all we can take, with a 30-inch wheel, 3-inch face and about 2-inch traverse per revolution. A work speed of 15 feet would give us about 3 revolutions per minute, making about 7 minutes for one cut across the roll and 14½ hours for the chilled portion of the roll.

For the soft necks of the roll (average reduction ½ inch) we can take a surface speed of 20 feet, equivalent to about 5 revolutions per minute, a feed of 0.004 inch and a traverse of about 2½ inches.

CHAPTER IV.

THE BURSTING OF EMERY WHEELS.

In 1902 some important tests of the strength of emery wheels were undertaken at the Case School of Applied Science, Cleveland, Ohio, under the direction of Prof. H. Benjamin. Fifteen wheels of various makes were tested to destruction. The results of these tests are given in the following.

Most manufacturers of this class of wheels test them for their own information, but the results are not generally given to the public. At the Norton Emery Wheel Works, all wheels are tested before leaving the shop at a speed double that allowed in regular service, and occasionally wheels are burst to determine the actual factor of safety.

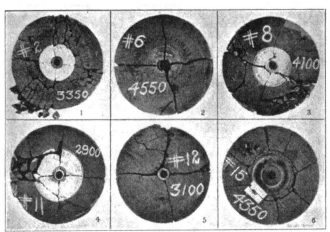

Fig. 13. Various Ways in which Emery Wheels Burst.

Emery-wheel accidents are not uncommon, but can usually be traced to the carelessness of the operator. One common cause of failure is allowing a small piece of work to slip or roll between the wheel and the rest.

The wheels selected for the experiments were all of the same size, being sixteen inches in diameter by one inch thick, and having a hole one and one-half inch in diameter. The object of the experiment being to determine the bursting speed of such wheels as are actually on the market, emery wheels were obtained through various outside parties without indicating to the agents or manufacturers the use to be made of them. In this way wheels of six different makes were obtained, the label on each wheel showing usually the maker's name, the grade number or letter, the quality of emery, and the speed recommended for use. As shown in Table IV, giving the results, the working speed varied in the different wheels from 1,150 to 1,400 revolutions per minute, the average being about 1,200 revolutions per minute. For a diameter of sixteen inches this corresponds to a peripheral velocity of about 5,000 feet per minute. The table also shows that the fineness of the emery varied from ten to sixty, the average being about thirty.

The wheels were held between two collars, each six and one-eighth inches in diameter and concaved, so as to bear only on a ring three-fourths of an inch wide at the outer circumference.

Table IV shows the results of the experiments in detail, and needs

TABLE IV. RESULTS OF TEST ON EMERY WHEELS.

No of Test.	Grade Mark.	No. of Emery.	WORKING SPEED.		BURSTING SPEED.		Speed Ratio	Factor of Safety.
			Revs. per Minute.	Feet per Minute.	Revs per Minute.	Feet per Minute.		
1	4 5	20	1,200	5,030	3,100	13,000	2.58	6.67
2	4.5	20	1,200	5,030	3,200	13,400	2.67	7.14
3	4.5	20	1,200	5,030	3,350	14,020	2.79	7.78
4	Q	30	1,250	5,230	3,750	15,700	3 00	9.00
5	Q	30	1,250	5,230	2 750	11,500	2.20	4.84
6	H	30	1,400	5,870	4 550	19,050	3.25	10.56
7	H	30	1,400	5,870	4,600	19,200	3.28	10 76
8	O	36	1,250	5,230	4,100	17,200	3.28	10.76
9	O	36	1,250	5,230	4,125	17,250	3.30	10.89
10	2 5	60	1,150	4,830	2,750	11,500	2.39	5.71
11	2.5	60	1,150	4,830	2,900	12,100	2.52	6.35
12	M. H.	14	1,200	5,030	3,100	12,970	2.58	6.66
13	.	24	1,200	5,030	3,800	15,900	3.17	10.00
14	H	10-12	1,200	5,030	4,100	17,200	3.42	11.70
15	H	10-12	1,200	5,030	4,350	18,200	3.62	13.10

Tests 6 and 7; wheels made with wire netting; tests 14 and 15, with vulcanized rubber.

but little explanation. The illustrations in Fig. 13 show characteristic fractures, and the appearance of various wheels after bursting. Wheels numbered 1, 2, and 3 in table were of one make, and show a remarkable uniformity in strength. Nos. 4, 5, 8, and 9 were all made by one firm; the two latter wheels were of finer grain than the others, and show a correspondingly greater strength. Nos. 6 and 7 contained a layer of brass wire netting imbedded in the emery, and were about one-third stronger than the average of the ordinary wheels. The wheels numbered 10 and 11 were the weakest among those tested, but have an apparent factor of safety of between five and six. Nos. 12 and 13, of still another make, burst at about the average speed. Wheels Nos. 14 and 15 were so-called vulcanized wheels, containing rubber in the bond, and intended for particularly severe service. These showed, as was expected, rather more than the average strength.

An examination of the last two columns in the table shows that the wheels burst at speeds varying from two and one-quarter to three and three-quarters the working speed, and accordingly had factors of safety varying from five to thirteen.

It is then apparent that any of these wheels were safe. at the speed recommended, and would not have burst under ordinary conditions. At the same time, considering the violent nature of the service and the shocks to which they are exposed, it would seem that the factor of safety for emery wheels should be large. In comparison with those generally used in machines, a factor of eight or ten would seem small enough. It may also be said that such a variation in strength between wheels of the same make and grade, as for instance, that between Nos. 4 and 5, indicates a lack of uniformity which causes distrust. The fractures were in the main radial, as may be seen from Fig. 13, the wheel splitting in three, four or five sectors as might chance. It may be assumed that these radial cracks started from the rim where the velocity and stress were greatest, but it is a fact

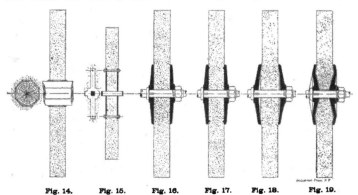

Fig. 14.   Fig. 15.   Fig. 16.   Fig. 17.   Fig. 18.   Fig. 19.

worthy of notice that in nearly every instance the cracks radiated from points where the lead bushing projected into the body of the wheel.

Fastenings for Emery Wheels and Grindstones.

When an emery wheel or grindstone is revolved, a certain amount of tension is set up and the method of fastening should be carefully considered, as the bursting of emery wheels may be accelerated by improper fastenings. The oldest fastening for this class of tool is that shown in Fig. 14. The stone has a large hole in its center and is carried by a shaft of the same type as that used on old-fashioned water wheels. It is fastened with wooden wedges, which are soaked with water in order to make them swell and tighten as much as possible. The instability of these wedges, which, by alternate wetting and drying, are constantly varying the tension put upon the stone, serves to increase the danger of the bursting of the same. This style of fastening is, therefore, seldom used at present.

A step in advance was made by the introduction of iron shafts. At first these were square in section. The stone was fastened between two wrought-iron crosses having a square hole at their centers. The ends of the arms were provided with holes through which bolts were passed for the fastening of the stone, as shown in Fig. 15. The crosses were themselves fastened to the shaft by wrought-iron wedges, so that the stone was, in this way, freed from all tension due to its fastenings. It was merely weakened in section by letting the crosses into it and by the holes for the bolts.

An essential improvement was made in the fastenings by the introduction of round shafts as shown in Fig. 16. Here the stone is fastened between two round clamping plates which are drawn together by a nut on the shaft. Through the pressing of these plates the stone is, of course, subjected to a crushing stress; it must, therefore, be admitted that stresses are thus set up in it that extend beyond the outer diameter of the plates. As a matter of fact stones so fastened have been sprung to such an extent that all of the material outside the plates has been fractured. Such clamping plates have also been made with circular ribs, as shown in Fig. 17, thus forming a first-class bursting furrow in the stone. These ribs possess the disadvantage that the section of the stone is dangerously weakened by them.

Another form of fastening is that shown in Fig. 18. Here cone-shaped plates are used for clamping the stone. The hollow cones are brought to bear against it. When these come to a bearing the plate must press equally against the stone throughout its whole circumference. As they are pressed together they are distorted to a greater or less extent on account of their own elasticity; their surfaces will be forced back and the diameter of the rims increased. Through the grinding action on the edge of the stone, the latter will increase in size and this increase will be shared by the plates which will thus set up a radial stress. These cone-shaped plates which seem to be so advantageous are, therefore, detrimental in that they exert a destructive influence on the stone.

The fastening by means of plates in the form of inverted cones, as shown in Fig. 19, is a preferable one. Here, by a tightening of the plates, the rim is drawn in towards the center and a tension towards that point is created. The tensions which are produced by the pressing of the plates together are toward the outside, and counteract each other to a great extent, so that there are no unfavorable stresses set up in the stone by this method of fastening. It will also be of advantage to make the bearing surface of the tightening nut spherical, whereby the plates can be made to better adjust themselves to any inequalities in the stone.

CHAPTER V.

GRINDING KINKS AND EXAMPLES OF GRINDING.

Grinding a Large Crank-shaft.

A prominent English chainmaker recently sent to the Norton Grinding Co., Worcester, Mass., a rough-turned crankshaft to be ground to the dimensions given in Fig. 20. The conditions given were that the throw must be ½ inch plus or minus 0.001 inch and that the keyway shown in Fig. 20 should line up exactly with the highest point of the eccentric. The keyway was already in the shaft, when received. The following method was pursued in preparing the crankshaft for the grinder:

Two cast-iron blocks, Fig. 22, were planed to the dimensions given, and one side, E in Fig. 23, was scraped to a surface-plate. A squaring chip was then taken across a lathe face-plate and the plate was rigged with blocks and parallels as per Fig. 23. The surface E of the parallel B was also scraped to a surface-plate. When the large hole was bored, the block A, Fig. 23, was against parallel C and when the small hole, or eccentric hole, was bored, A was moved along parallel B and block D was inserted. Tissue paper was used in both settings to insure actual contact. The large holes were bored 0.015 inch larger than the finished diameter of the crankshaft ends. After boring the small holes, a 1-inch arbor was forced into the small holes and the 60-degree center holes were turned with a lathe tool. The truth of these 60-degree holes was tested by means of a ground cone point and red lead. A tapped hole and setscrew completed each block.

The shaft was now prepared for the blocks by grinding each end a wringing fit for its block. Before doing this, the center holes in the shaft were tested and scraped to a 60-degree cone point, to insure a perfectly round shaft when ground.

The next operation was to correctly locate the keyway. For this, two blocks, A and B, Fig. 21, were made. A is a 1-inch block that tapped lightly into the keyway and projected a short distance, as shown. B is a block planed to micrometer gage, and of such a height as to bring the center line of the keyway and the center line of the crankshaft into a plane parallel to the planer surface C, Fig. 21. The proper height of B was easily found by means of micrometer measurements and deductions. Having made A and B, Fig. 21, the whole job was taken to a newly planed planer table and the end blocks were placed on the crankshaft. A was then placed in the keyway and the crankshaft turned until A rested on B. With tissue paper under the end blocks D, Fig. 21, and between A and B, adjustments were made until all the papers held fast. The blocks D were then made secure by means of the setscrews E. After a final test with the tissue papers, the crank-

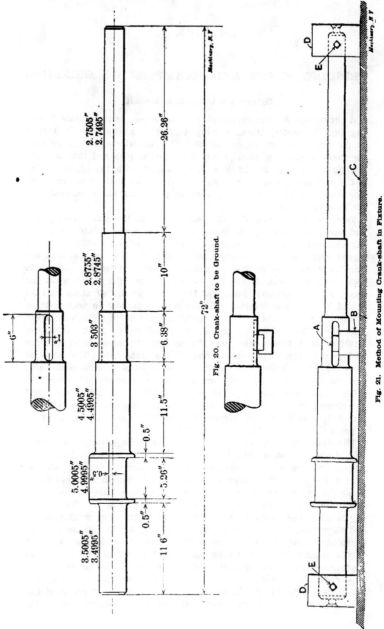

Fig. 20. Crank-shaft to be Ground.

Fig. 21. Method of Mounting Crank-shaft in Fixture.

shaft was ready to have the eccentric ground. This was done on an 18-inch by 96-inch Norton plain grinder. The fillets on the eccentric were also ground at the same time.

The length of throw was tested in the grinder by means of a Bath

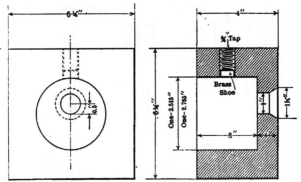

Fig. 22. Fixture for Grinding Crank-shaft.

indicator and a 1-inch B. & S. disk, and found to be within the required limits. When the eccentric was completed, the end blocks were

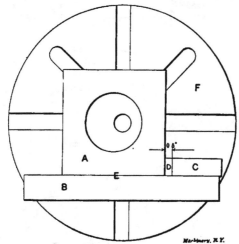

Machinery, N. Y.

Fig. 23. Method of Boring the Fixture Used for Grinding the Crank-shaft.

removed and the remainder of the crankshaft was ground on its own centers.

Grinding Kinks.

In the following are described some of the kinks used by toolmakers in grinding; these kinks were contributed to MACHINERY by Paul W. Abbott.

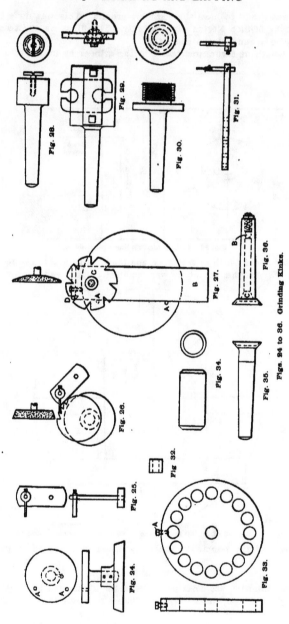

Fig. 28. Fig. 29. Fig. 30. Fig. 31.

Fig. 27. Fig. 36.

Fig. 34. Fig. 35.

Figs. 24 to 36. Grinding Kinks.

Fig. 26.

Fig. 32.

Fig. 25.

Fig. 33.

Fig. 24.

Fig. 24 is a hand grinding rest which is very handy for use on the universal grinder. It is adjustable up and down for height, and is used for hand grinding circular and straight form tools, sharpening metal slotting saws, formed cutters, etc. Fig. 26 shows the application of the hand rest to the grinding of saw teeth in a blank. The tooth rest used in connection with this operation is shown in Fig. 25. These saws are first ground on an arbor, the old teeth being ground off, leaving a perfect circle. The operator then puts on this device. setting the tooth rest so that the teeth will be about ¼ inch apart, and grinds around by hand, not quite bringing each tooth to a sharp point. On the last nine or ten teeth he evens up any inaccuracy in the spacing, the wheel being trued off to the exact shape of tooth space wanted.

Fig. 27 shows a device for accurately sharpening formed cutters up to 3 inches diameter, which is used when the cutter grinder has another job in it, or could be used to advantage where there was no surface or cutter grinder. The device consists of the cast iron slide *B*, at the end of which is a tapped hole *C*, with a small fillister head screw which holds the various sizes of bushings which fit the holes in the cutters. On the same end is the index pin *D*, which is adjustable back and forth. In operation, the hand rest shown in Fig. 1 is also used, and the pins *A* are lined up parallel with the forward travel of the wheel and so that the cutting face of the wheel is on a line with the center of the bushing. The cutter is then slipped on over the bushing and the index pin is set so that the required amount will be ground from the face of the tooth. The operator brings the wheel up to the proper position and then pushes the slide forward until the wheel has reach the bottom of the tooth space; he then withdraws the slide and indexes to the next tooth, and so on, tooth after tooth. It will be noticed that the index pin rests against the back of the tooth, which means that upon the previous milling of the teeth depends the accuracy of the grinding; but on the standard cutters furnished by numerous concerns this spacing will be found accurate enough.

Fig. 28 is a center for the head-stock for holding small forming tools of odd size, or threaded pieces which are to be ground on the periphery. The tools are simply clamped to the face of the center, and trued up by an indicator. Fig. 29 is a device for the tool grinder for grinding snap gages, where there is no surface grinder for this class of work. The shank of this device is made to fit the head-stock, and the gages are clamped to it by a small strap and two screws. This fixture revolves while in use, and the jaws of the gage are ground by feeding a thin wheel in and out by hand. Revolving the device insures perfectly straight gage faces.. Fig. 30 shows a center for the universal grinder for holding a standard line of large end milling cutters with threaded holes, while sharpening. The head-stock is swung around at right angles to the ways, and with a long support for the tooth rest (Fig. 31), which is bolted to the platen, the cutters are ground very handily by throwing in the feed and grinding one tooth, and then, before the wheel comes back, indexing to the next tooth, and so on.

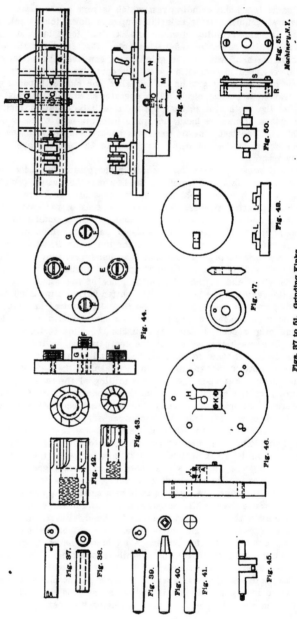

Figs. 37 to 51. Grinding Kinks.

Fig. 32 shows a hardened roller which is ground all over, and Fig. 33 the fixture for the universal grinder for grinding the sides of this roll. This plate was made of cast iron, with both sides ground and with each hole ground to 0.0005 inch over standard size. Each hole has a ¼-inch set-screw, as shown at *A*. In operation, the plate is fastened to the face-plate by a draw-back rod, and the head-stock is swung around at right angles. As the plate revolves, 16 rolls are ground at once, first on one side, and then the plate is turned and the other side ground, the rolls being made to standard length by using a depth gage. The hardened roll shown in Fig. 34, which is used on swaging machines, is held by the centers shown in Fig. 35 and 36, when being ground. Fig. 35 is the head-stock center cupped out on the end to fit the beveled end of the roll. This center drives the roll by friction, the pressure being obtained by the spring tail-stock. Fig. 36 is the tail-center which is in two parts, the inner spindle running with the roll and being adjusted by the screw in the end so that the thrust is taken by the ball *B*, the tapered portions *C* just clearing each other. Other methods of grinding rolls are shown in Figs. 37 to 41. One example of grinding is shown in Fig. 37, and its center in Fig. 39. The roll is driven by a pin on the center, which engages with a corresponding hole in the work. A better method is to center the roll and then in one end drive a square 60-degree punch, using the square center shown in Fig. 41 for driving the work while grinding. Another good method for hollow rolls, such a shown in Fig. 38, is to use a 15-degree square center, such as shown in Fig. 40, the end of which just enters the hole.

Figs. 42 and 43 show two end mills. The smaller one is fastened inside of the larger when in use, and when in position rests against the bottom of the hole and projects outside a definite distance. The length *D* is standard in all these mills. Fig. 44 shows the fixture for grinding two pairs of these mills at a time, so that the same amount will be taken off of both the short and long ones. Threaded bushings *E* fit the larger size mills, and *F*, the smaller. The collars *G* are of such thickness that the cutting face of the smaller mill is brought into the same plane as the larger, and so when grinding, an equal amount is removed from the face of each mill. The plate is held to the face-plate by a draw-back rod. The head-stock is swung at right angles, and with the fixture revolving, the wheel traverses back and forth across the faces of the mills. The mills are then taken to a cutter grinder and backed off.

Fig. 45 shows a small crank-shaft, and Fig. 46 the fixture for grinding the pin. The bearings are first ground on centers in the usual way. The fixture is of cast iron and is held to the face-plate by screws and dowel pins. In the making of this fixture the hole *H* was ground out to the size of the bearing, and then the fixture was correctly located and doweled to the regular face-plate. The crank, while being ground, is held by the set-screws *J*, and the screws *K* which are set against the crank on either side.

The grinding of formed cutters, similar to the one shown in Fig. 47, so that they will be interchangeable, is very interesting. The error limit is 0.00025 inch. The grinder used is a Norton universal tool and cutter grinder. After hardening, the cutters are first ground to a definite thickness. For this operation they are held against the face-plate by a draw-back chuck. The next operation is grinding the bev-eled sides, which is accomplished by holding the cutters against a small face-plate by a draw-back chuck. The correct angle of bevel is obtained with the protractor, and to get the correct diameter of the bevel sides, and to insure that the bevel sides stand exactly in the same relation to each other, the gage shown in Fig. 48 is used. This gage is hardened and ground all over, and the two gaging points L are set a predetermined distance apart and as near the same height from the platen as mechanical means can make them. It is obvious that cutters which are all ground the same thickness, and which will pass through this gage with the beveled sides both touching the gage points with equal pressure, will interchange within pretty close limits. The operator grinds one bevel side at a time, trying the work every little while in this gage; when one side passes through the gage the cutter is turned around and the other bevel ground. For grinding the radius on the periphery and bringing the cutter to the correct diameter, the radius grinding fixture shown in Fig. 49 is used. The dovetailed base M is fitted to the platen of the grinder and upon this base is a slid-ing base N which is pivoted to M by a bolt O. Upon the base N there is an auxiliary platen P which can be adjusted back and forth by the screw Q for getting the proper radius. This auxiliary platen is made the same as the machine platen so that the regular head-and tail-stocks will go on it. A cutter is placed on a special arbor and the platen P adjusted to give the correct radius. The wheel is then brought up and the cutter is ground to the correct diameter, the curved face being obtained by swinging the base N back and forth by hand in an arc of a circle, with bolt O as a center.

Another ingenious scheme is shown in Fig. 51. Three or four pieces similar to the one shown in Fig. 50 were to have the holes ground out. With an independent 4-jawed chuck this would have been easy, but there was no such chuck; and as there would never be any more of these pieces to be ground the fixture for doing the work had to be inexpensive. The face-plate could not be used as the pieces were smaller than the hole in the face-plate. The operator thought awhile, and then hunted around a few minutes and found a large washer R, tapped two holes in it, filed up the sheet steel strap S, and with a couple of machine screws was ready to begin. The washer was first put in the universal chuck and the outer side ground. One of the pieces was then clamped in place, and after putting on the internal grinding attachment it was ready to be ground.

Grinding Fixture for End Measure Rods.

The fixture shown in Fig. 52 was made to grind spherical end meas-uring rods 3 inches long, and over. There were three diameters of these rods, ⅜, ½ and ⅝ inch, and three sizes of fixtures were also made to suit. Steel sleeves *B* were made of various lengths for each fixture, to stiffen the rods near the end when grinding. These sleeves had a bearing at each end in hardened bushings which were driven into the fixture body. The sleeves were kept from turning and from sliding endways, independently of the gear, by the set-screws shown. These screws, in addition, held the rod being ground. Member *A* was a tight fit in the large bevel gear, which was made separate to allow of cutting the teeth. The flush oilers *C* were placed on opposite sides of the body, so the tendency to throw oil would be done away with as much as possible.

A feature apt to be overlooked is that the number of teeth in the two gears must be prime to each other. If not, every certain number of revolutions, depending on the gears used, the wheel will grind into

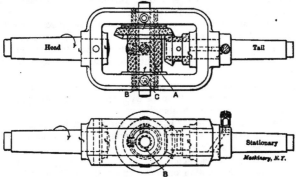

Fig. 52. Fixture for Grinding Spherical End Measure Rods.

one of the previous cuts. The fixtures were arranged to be used on a bench-lathe, with a tool-post grinding fixture.

Selection of Wheel.

The following little hints regarding grinding, taken from a booklet issued by the Norton Co., Worcester, Mass., will prove of value to all who have to do with grinding machines and grinding.

Don't believe that all materials can be ground equally well with one and the same wheel.

Get the proper wheel for the work.

You would not expect to turn all kinds of lathe work with one tool having only one form of cutting edge. The grinding wheel is a tool for cutting.

Different shapes of work, different kinds of metal, require different cutting edges as well when grinding as when turning. Different grades and grains of wheels are required for different kinds of work.

Grinding wheels are numbered from coarse to fine, and graded from soft to hard. The grade is denoted by the letters of the alphabet from E to Z.

Don't decide on the wheel without knowing the work.

Spindle speed and character of the material, shape of work to be ground, and surface of wheel in contact are prime factors.

In cylindrical grinding, speed of work, diameter of work and depth of cut must all be reckoned with in the selection of the right combination of grain and grade.

The condition of the machine affects the efficiency of the wheel. Heavy machines with large wheel spindles and massive wheel support call for a wheel different from those for lighter machines with smaller spindles.

Don't order a certain grade wheel merely because that grade is used on similar work in another plant.

Don't use a hard wheel to economize—it is production you are after.

A hard wheel is more likely to change the temperature of the work or to become glazed than a soft one; furthermore, it requires more power to do the same amount of work.

It is a common error to assume that a wheel for grinding steel and cast iron, chilled iron and hardened steel, must be as fine as the surface desired. A coarse wheel will produce a fine finish if the proper relations between grade, depth of cut, speed of work, speed of wheel, etc., are observed.

When grinding brass and the softer bronzes, the wheel must be as fine as the finish required. Bronzes with "manganese" or "phosphor" permit the use of coarser wheels.

Don't get a wheel made for soft steel for use on hard steel.

For a fine finish on hard stock, a coarse wheel may be necessary, and the harder the stock the coarser the wheel.

When ordering wheels, don't forget the diameter, width, style of face, arbor holes, description of work, speed of spindle, and the number and letter denoting the combination of grain and grade, if known.

The width of the wheel should be in proportion to the amount of the material to be removed with each revolution of the work.

If you reduce the width of the wheel, you must use a finer feed, and consequently do less work.

Mounting.

Never mount wheels without flanges.

Flanges should be, at least, one-third the diameter of the wheel; one-half is recommended. Flanges should be concave—never straight or convex.

Use fiber or rubber washers a trifle larger than the diameter of the flanges, or flanges with soft metal facings.

Hooded machines are desirable when practicable.

Truing.

Don't start work on a new wheel until you are sure it runs true.

Always have a wheel dresser handy for truing wheels for off-hand grinding.

Never use a dresser on wheels that grind circular work on centers.

For truing wheels used on plain cylindrical and universal grinding machines, cutter and reamer grinders, etc., the diamond is recommended. To obtain the best results it is absolutely necessary.

Never attempt to true a wheel for circular grinding unless the diamond is held in a rigid tool-post on the table of the machine. You cannot do good work with such a wheel when it is trued "by hand."

To get a truly ground surface, you must keep the face of the wheel true.

The quality of surface finish is dependent on the conditions of the wheel face and depth of cut.

Speed.

Don't start grinding until you know the speed is right—not "near enough," but right.

Even a slight variation in speed may be the cause of success or failure of any wheel.

Failure is sometimes turned into success by merely changing the speed of either the wheel or work.

Speed up the spindle as the diameter of the wheel is decreased. Approximately the same peripheral rate should be maintained as the wheel wears down.

Complaint is sometimes made that wheels appear to be softer toward the center. Usually this is because the same surface rate of speed is not maintained as the wheel is reduced in diameter. This causes the wheel to wear away faster and appear softer. It is also true that while the grade of the wheel may be uniform throughout, yet the smaller line of contact due to the smaller diameter will cause the wheel to appear softer.

Increasing the speed of a grinding wheel gives the effect of a harder wheel; decreasing the speed gives the effect of a softer wheel.

For surface grinding, it is customary to run wheels at a somewhat slower rate of speed than for general grinding. A speed of 4,000 to 5,000 surface feet is usually employed.

Wheels are run in actual practice from 4,000 to 6,000 feet per minute.

General Suggestions.

Transferring a wheel worn down to a small diameter, from a large machine to a small one, is good practice.

Keep the tickets or tags which are sent on the wheels in a record book, so that if a wheel is not satisfactory, reference can be made to order number when making complaint. It is equally valuable as a reference when ordering duplicate wheels.

Don't use the wrong wheel on a job because it will require a few minutes time to change wheels. A stop watch will prove to you that changing wheels is cheaper.

There is seldom a case where one and the same wheel can be used on all work without a greater loss of time than the change of wheel would involve. Many times, the time saved in grinding a single piece more than pays for changing the wheel.

Considerable difference in diameters of work will affect the cutting quality of a wheel on any given material.

A successful wheel on the small diameters may work much slower on the larger diameters.

The wheel most suitable for work of very large diameter may wear away too fast on work of smaller diameter.

A suitable wheel for small diameters may cause chatter on pieces of large diameters.

Don't grind circular work dry.

A good wheel will grind in water, soda water or oil.

Water keeps the wheel working cool, and increases grinding production.

Soda water keeps the work and the machine from rusting.

Oil in soda water increases the wheel's effectiveness.

The particles from a grinding wheel do not adhere to steel. Don't let any one convince you to the contrary.

Grinding is profitable for removing stock as well as for finishing.

Keep the face of the wheel true and parallel with axis of spindle. Vibration makes grinding wheels wear.

Keep all rests adjusted close to the wheel, otherwise work is liable to be caught and injury result.

Keep boxes well oiled and adjusted.

When practicable, indicate on each machine the revolution of spindle and size of wheel to be run upon it.

Don't disregard the setting up instructions that go with the grinding machine.

CHAPTER VI.

LAPPING FLAT WORK AND GAGES.

The main essential points of the art of lapping can be described in a book, but, the same as with any other line of mechanical work, it is necessary that the workman shall do considerable lapping before he can become proficient. There are certain motions, touches, sounds, refinements, etc., which the skilled workman acquires by practice, that are impossible of enumeration and description, or of enumeration and description that would be intelligible to an inexperienced man. For instance, ask a carpenter how he knows that he is sawing a board straight, and he will be unable to tell you. Nevertheless, he has acquired a peculiar sense of touch, or such general acuteness

Fig. 53. Back of Standard Flat Lap, showing Ribbed Construction.

of the senses, that he knows instantly when the saw starts to "run out." His mind and arm automatically, as it were, return the saw to a straight line without missing a stroke. It is the same way with a die maker. He can file a die, looking only at the surface line, and can detect the instant when his file "rocks" from a straight line. He will tell you that he "feels" it, but is unable to define what the sensation is. Likewise, one cannot explain some of the finer points in the art of lapping, and can only point out those which are fundamental, and which must be acquired first by the workman unaccustomed to such operations. He must acquire the refinements by practice and experience.

A Perfect Lap Required for Perfect Lapping.

The first requisite of perfect lapping is a perfect lap, and right here is where the novice will make his first mistake, that is, in the preparation of the lap. To make a surface lap, it should be carefully planed, strains due to clamps being avoided, and then it should be carefully scraped to a standard surface plate. This is done by rubbing the face of the lap on the standard surface plate and scraping down the high spots until a perfect plane surface is obtained. If a standard surface plate is not at hand, a lap can be made level by using three laps that are nicely planed and used alternately as follows. We will number the laps Nos. 1, 2, and 3. Now, rub No. 1

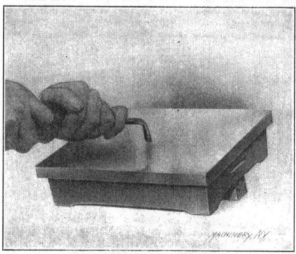

Fig. 54. Scraping Down the High Spots.

and No. 2 together, and scrape the high spots until they fit. Then introduce No. 3 and scrape it down to fit No. 2, and then to fit No. 1, and so on. The third lap eliminates the error that might follow if only two laps were used. For example, it is possible to fit two plates accurately together without making them plane surfaces, one becoming concave and the other convex. The third lap absolutely prevents this and produces a perfect plane surface if time and patience hold out. It is a slow operation, but not so slow as trying to lap a piece true with a lap that is not true.

The Objection to Ground Laps.

The laps may be ground together instead of scraping, but the writer prefers the scraping process, as it is easy to see when the job is done.

It is also better to scrape them, because it is quicker than attempting to grind them level with the fine grade of emery that is required for nice lapping, and it must be remembered that when ground together the laps *are already charged*. Hence, the necessity of using a fine grade of emery if they are ground together.

Using a Hand Surface Lap.

The writer prefers a cast iron lap, Fig. 53, thoroughly charged, and having all loose emery washed off with gasoline. When lapping, the surface is preferably kept moist with kerosene, although gasoline causes the lap to cut a trifle faster. It evaporates so rapidly, however, that the lap soon becomes dry, and the surface caked and glossy in

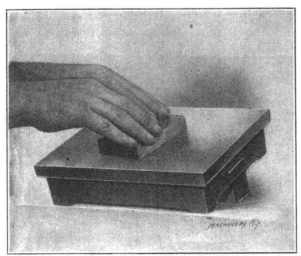

Fig. 55. Charging the Lap, using a Hardened Steel Block.

spots. When in this condition, a lap will not produce true work. The lap should be employed so as to utilize every available part of its surface. Gently push the work all around on its surface, and try not to make two consecutive trips over the same place on the lap.

Do not add a fresh supply of loose emery to a lap, as is frequently done, because the work will roll around on these small particles, which will keep it from good contact with the lap, causing poor results to follow. If a lap is thoroughly charged at the beginning, and is not crowded too hard, and is kept well moistened, it will carry all the abrasive that is required for a long time. This is evident, upon reflection, for if a lap is completely charged to begin with, no more emery can be forced into it. The pressure on the work should only be sufficient to insure constant contact. The lap can be made to cut only so fast, and if excessive pressure is applied, it will become "stripped" in places, which means that the emery which was imbedded in the lap has become dislodged, thus making an uneven surface on the lap.

Causes of Scratches—Grading Emery.

The causes for scratches are as follows: Loose emery on the lap; too much pressure on the work which dislodges the charged emery; and what is, perhaps, the greatest cause, poorly graded emery. To produce a surface having a high polish free from scratches, the lap should be charged with emery or other abrasive that is very fine. The so-called "wash flour emery," sold commercially, is generally too uneven in grade. It is advisable for those who have considerable high class lapping to do to grade their own emery in the following manner: A quantity of flour emery is placed in a heavy cloth bag, and

Fig. 56. Lapping a Gage.

the bag gently tapped. The finest emery will work through first, and should be caught on a piece of paper. When sufficient emery is thus obtained it is placed in a dish of lard or sperm oil. The largest particles of emery will rapidly sink to the bottom, and in about one hour the oil should be poured into another dish, care being exercised that the sediment at the bottom of the dish is not disturbed. The oil is now allowed to stand for several hours, say over night, and then is decanted again, and so on, until the desired grade of abrasive is obtained.

For the information of those not well acquainted with grading abrasives, it may be said that the grade of diamond dust known as "ungraded" is obtained in about five minutes, while it requires about three weeks to obtain the grade known as No. 5, which is very fine. But, even at the end of three weeks there still will be small particles in the oil that have not settled, due to the viscosity of the oil.

To lap true and free from scratches, one must have skill and be thoroughly conversant with the peculiar sounds, touches, and motions spoken of above. For a high polish on work, a rapid motion and slight pressure are necessary for success. It is also necessary that the lap is properly charged with properly graded abrasive.

Lapping Gage Jaws.

Fig. 56 shows the best method that has come to the writer's notice for lapping the jaws of gages. The lap is made of cast iron and is relieved as shown, leaving only a thin edge or flange on each side to bear against the jaws. As the machine table is worked back and forth, the lap passes over the entire surface of the jaw, grinding it down in the same manner as would be done with a cup emery wheel.

Fig. 57. View of Machine Vise, showing a Gage Clamped without Springing it.

Care must be taken to clamp the gage in the vise so as not to spring it. Fig. 57 illustrates an approved method for holding a gage so that the vise jaws will not deflect it. Should the gage be sprung, it is clamped at the center only, leaving the ends free. Snap gages are now mostly made of machine steel and pack-hardened. Made in this way they do not change much, as the interior of the gage is left soft, and whatever change occurs can be easily remedied, but in any case the method illustrated is the safest one to follow, for it leaves no doubt as to the gage being held free from spring during the lapping operation.

A lap should be turned on the arbor on which it is to be used, for it is almost impossible to put a lap back on an arbor after it has been removed, and have it run true. Therefore, the lap should be recessed quite deeply, as shown, to allow for truing up each time the lap is placed on the arbor. Perhaps when the lap is mounted on an arbor in the milling machine, it will be found to run out not more than 0.001 inch, but that means that it is touching the work in only one spot, and the result can be hardly better than if a fly-cutter was used for a lap. Fig. 58 shows the operation of truing the lap. A keen cutting tool is clamped in the vise and in this way the lap can be trued as nicely as though it were done in the lathe. In fact, it is superior, for there is absolutely no change in the alignment of the lap with the work spindle after it is turned, which might easily happen should it be turned in the lathe and then mounted in the milling machine spindle. With a perfectly true lap, a perfect contact between the lap and gage is insured for its entire circumfer-

Fig. 58. Truing the Lap with a Tool held in the Vise.

ence. Both sides of the lap should be turned at the same setting on the arbor.

Fig. 59 shows the operation of charging a circular lap, using a roller mounted in a suitable handle for the purpose. The emery is rolled in under moderate pressure. It is good practice to make the roller of hardened steel, and after charging the lap, all the surface emery should be thoroughly washed off.

The next step is to square up the jaws of the gage. Do not depend on the zero marks of the vise. The jaws of the gage may have sprung a little in hardening, and if the zero marks of the vise are depended upon to square the work, there possibly will not be sufficient stock on the jaws to clean up. Be very careful to set the gage by the surface of the jaws and to clamp it in the vise as previously noted, so that it is under no pressure tending to spring it out of shape.

When employing a power-actuated lap, the little instrument shown in Fig. 60 is useful in determining the instant when the lap touches the work. By placing the forked end on the work and the wooden part to the ear, the sound is greatly magnified, and it makes it much easier to determine the precise point of initial contact. If one depends upon the naked ear to tell when the lap touches the work, he is liable to crowd the lap too much, and scratch the work or strip the lap. With this instrument the mechanic will know the instant the lap just touches the work, and this is the position where its work should be done. In short, the lap should not work under any appreciable pressure, but should simply touch the work. Hence the desirability of some means of magnifying the sound and not depending on the naked ear.

The workman should avoid the custom of adding a fresh supply

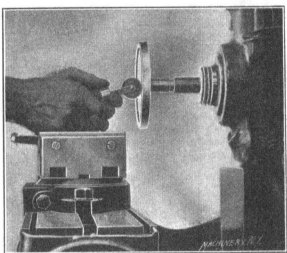

Fig. 59. Charging the Lap with a Roller.

of abrasive to the lap, as it is not only injurious to the character of the product, but it naturally increases the time required for lapping. To illustrate the action, suppose that an arbor is to be ground in a grinding machine, and that it is belted so that it runs with a wheel at the same speed. The consequence will be that no grinding action could take place, as there would be no difference in motion. The condition is very similar when loose emery is placed on a surface lap. The emery simply rolls around between the work and the wheel, and occasionally a piece of emery is imbedded in the lap long enough to scratch the work. While it may look as though the lap was cutting much faster, the truth is that it cuts slower and produces poor work.

In lapping jaws, some workmen round-lap, and then finish by hand, but a better job will result when finished in the machine. It is poor practice to rough-lap a gage, using a coarse grade of emery, and then wash the lap and smear it with fine emery. Of course the lap is already charged with a grade of emery last used, and the act of putting on a supply of fine emery on the lap will not produce as good a surface as if the gage were finished without the fresh supply of emery, though the latter is of a finer grade.

Lapping Gages.

Assume that a 1-inch plug gage is to be lapped to size. Such a gage needs only about 0.0015 inch for lapping. The outside lap, shown

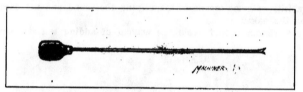

Fig. 60. Sound Magnifier.

in Fig. 62, should be made of cast iron, copper or lead, and the holder *D* should be provided with adjusting screws. The flour emery used should be sifted through a cloth bag to prevent any large particles of emery entering the lap and scratching the gage. After sifting the emery it is mixed with lard or sperm oil to the consistency of a thin paste. The gage is then gripped in the chuck of the lathe by the knurled end and smeared with emery paste. The lap is adjusted to fit snugly on the gage and the lathe is speeded up as fast as possible

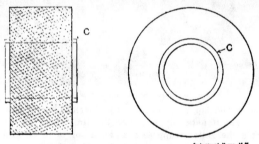

Industrial Press, N.Y.

Fig. 61. Ring Gage to be Lapped.

without causing the emery to leave the gage. The lap requires constant adjusting, to take up the wear of the lap, and reduction in size of the gage. When measuring the gage, it should be measured at both ends and in the center to make sure that it is not being lapped tapering. When the gage has been lapped to within 0.0002 inch of the finished size, allow the gage to thoroughly cool and then by hand lap lengthwise of gage to the finished size. By so doing all minute ridges that are caused by circular lapping are removed, thereby leaving a true surface and also imparting a silvery finish. A gage should never be lapped to size while warm (heated by the friction of the lap), because the gage expands when heated, and if then lapped to size it will contract enough to spoil it.

In grinding out the inside of a ring gage considerable difficulty is experienced in adjusting the grinder so that it will grind straight. One way to prove the straightness of a hole being ground is to move the wheel over to the opposite side of the hole until the wheel will

 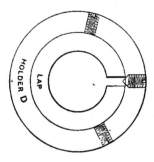

Fig. 62. Outside Lap.

just barely "spark." Then, beginning from the back of the hole, feed out, and if the hole is tapering, the wheel will either cease to spark, or will spark considerably more. Another and better way is by means of the multiplying indicator gage, Fig. 64. By fastening the indicator to the spindle of the grinder and placing the contact pin of the indicator on the opposite side of the hole and feeding in and out, the pointer will record in thousandths of an inch just what the deflection is.

A ring gage should be made as shown at Fig. 61, the object being

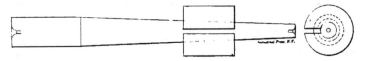

Fig. 63. Inside Lap.

to prevent the gage becoming "bell muzzled" while lapping. After the gage is finished, the thin projecting web *C* is ground off, leaving a good straigth hole. The lap used for inside work is shown at Fig. 63. The lap can be made to always fit the gage by merely forcing the lap further along on the taper arbor; the lap being slotted allows it to expand. In making a ring gage having a taper hole or a taper plug gage, it is necessary to employ a different method of lapping, as it is impossible to lap a taper hole with a taper lap. The facts regarding lapping are these: First, the lap must fit the hole at all times; secondly, the lap must constantly be moved back and forth. Therefore, if a taper lap is made to fit the taper hole it will lock

and not revolve. If held in one place the lap will quickly assume
the uneven surface of the hole. If the operator attempts to lap a
taper hole by constantly revolving the gage on a straight lap he will
surely dwell longer in one place than another, thereby making a hole
that is anything but round. The following method is, therefore,

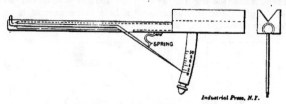

Fig. 64. Indicator for Testing Truth of Holes.

used: Having ground the hole to size, plus allowance for lapping,
then, without disturbing position of slide rest or grinder head, change
the emery wheel for a lap made of copper—of the same shape as the
emery wheel with the exception of having a wider face—and lap in
the same manner as the hole was ground, care being taken not to
"crowd" the lap.

CHAPTER VII.

THE ROTARY LAP.

In Fig. 65 is shown a rotary lap 24 inches in diameter, intended to
run at a speed about 300 revolutions per minute. The engraving will
give a clear idea as to the construction. Some men think that if the
lap has a true, flat surface, any one can produce true work, but such
is not the case; it requires considerable skill, and that skill can be
acquired only by long practice. Many machine operations can be
shown to another person and the principle grasped readily, but not
so with lapping. A great deal of skill is required in lapping thin
pieces, small straight-edges or long narrow bars. It is possible, and
requires no skill at all, to lap a thin piece of steel, convex or concave, by using a little more pressure in one place than another, and
if the surface of the lap is not kept sharp it will soon heat and warp
the work out of true. For a rotary lap kerosene and gasoline used
together give the best results; but a hand lap should always be used
dry. Keeping the surface of the rotary lap straight and true is very
important and requires good judgment in using it. The outer edge
runs so much faster than the inside that it is obvious that if it is
used too much, either in the middle, inside, or outside, hollow places
will be worn in the surface, so that it is a good plan to use the lap

all over and try the surface frequently with a straight-edge, favoring the high places when using it.

The rotary lap is charged by sprinkling carborundum over the surface when not in motion, and then pressing it in by rubbing with a piece of round iron held in both hands. An old pepper box with a perforated top is just the thing to use for sprinkling carborundum.

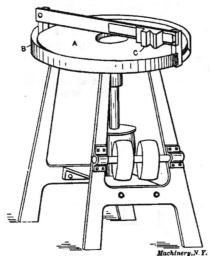

Fig. 65. The Rotary Lap.

The lap in Fig. 65 is made of an ordinary cast iron disk *A* with ribs on the bottom and anchor grooves on the top. It is also provided with a hub and a shaft. The end of the shaft supports the whole weight of the lap and runs on a hardened convex disk. A coating

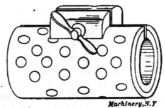

Fig. 66. Lap for Cylindrical Work.

of lead is cast over the surface of the lap, and then hammered to make it compact. A galvanized iron pan *B* is provided, the edges of which project above the surface of the lap to prevent the liquid, or whatever is used, from flying off, and onto everything around. Another handy device on this lap is a bar which is provided with ways and a sliding head *C* which can be pushed from the outer edge to the center of the lap. The bar is fastened to lugs which project

on opposite sides of the frame, and can readily be removed when not
in use. The bar is also provided with adjusting screws to set it
parallel with the surface of the lap or to set the sliding head square
with the surface. The sliding head has a square corner and an angle
groove which can be used for lapping the ends of round or square
pieces.

The engraving Fig. 66 shows a good way of making a lap for cylindrical work in the lathe. A piece of wrought iron pipe about 2 inches
long, with a number of 5/16-inch holes drilled through will answer
the purpose. Face one end of the pipe so that it will stand level on
a surface plate; wrap a piece of heavy paper around the outside, using
rubber bands to hold it on, then form a lug out of pasteboard and
clamp that on any side of the pipe so that several of the holes open
into it; secure a mandrel the same size as the piece to be lapped,

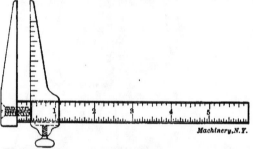

Fig. 67. Tool Used for Lapping when Sizing Duplicate Parts.

twist a piece of string in a spiral around the mandrel, insert it in
the center of the pipe, and pour the molten lead in. When cool, drive
out the mandrel and proceed to drill and tap a hole for a thumb-
screw in the center of the lug; then slit one side through the center
of the lug with a hack saw, and file off all sharp edges and burrs so
that it will not injure the hands. In lapping, the faster the lap is
drawn back and forth over the work the more nearly straight it will
be, and as the lap wears and works easy, the thumb-screw is given a
slight turn to keep it in contact with the work.

A very handy tool, which is shown in Fig. 67, is almost indispensable in doing work with the rotary and hand laps; it is a home-made
caliper square. The taper between the hardened jaws is 0.001 inch
in the whole length, and a 6-inch flexible scale is inserted in the
beam. One jaw is graduated, which enables one to see how far the
piece will slide up in the jaws. This caliper is not used so much
for accurate measurements as for accurate sizing for parallelism or
duplicating sizes.

My26C

18 Jan '40

5 Jul'4#

Printed in the USA
CPSIA information can be obtained
at www.ICGtesting.com
LVHW051037151123
763818LV00005B/453